Small Signal
Microwave Amplifier Design:
Solutions

Small Signal Microwave Amplifier Design: Solutions

Theodore Grosch

Noble Publishing Corporation
Atlanta, GA

NOBLE
PUBLISHING

Printed in the Unites States of America

To order contact:

Noble Publishing Corporation
4772 Stone Drive
Tucker, GA 30084

Phone: 770-908-2320
Fax: 770-939-0157
E-Mail: orders@noblepub.com
Internet: www.noblepub.com

Library of Congress Catalog Card Number: 99-067616

ISBN 1-884932-09-6

Contents

Solutions

2.1

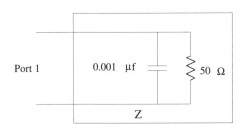

$$Z = \frac{V}{I}$$

$$I = I_R + I_C = \frac{V}{50} + \frac{V}{\left(\dfrac{1}{sC}\right)} = \frac{V}{50} + Vj\omega(1.0\text{E}-9)$$

$$Z = \frac{V}{\dfrac{V}{50} + Vj\omega(1.0\text{E}-9)}\ \Omega = \frac{50}{1 + j\omega(5.0\text{E}-8)}\ \Omega$$

2.2

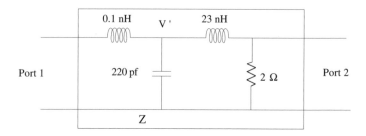

$$I_1 = I_c + I_r$$

$$I_r = \frac{V'}{sL_2 + R}$$

$$I_c = V'sC$$

$$V_{L_1} = V_1 - V' = sL_1 I_1 = sL_1 \left(\frac{V'}{sL_2 + R} + V'sC \right)$$

$$V_1 = sL_1 \left(\frac{V'}{sL_2 + R} + V'sC \right) + V'$$

$$Z_{11} = \frac{V_1}{I_1}\bigg|_{I_2=0} = \frac{sL_1 \left(\dfrac{1}{sL_2 + R} + sC \right) + 1}{\dfrac{1}{sL_2 + R} + sC} = \frac{sL_1 \left(1 + sC(sL_2 + R)\right) + sL_2 + R}{1 + sC(sL_2 + R)}$$

$$Z_{11} = \frac{s^3 L_1 L_2 C + s^2 CRL_1 + s(L_1 + L_2) + R}{1 + sCR + s^2 CL_2}$$

$$Z_{11} = \frac{R - \omega^2 CRL_1 + j\left(\omega(L_1 + L_2) - \omega^3 L_1 L_2 C\right)}{1 - \omega^2 CL_2 + j\omega CR}$$

$$Z_{11} = \frac{2 - \omega^2(4.40\mathrm{E}-20) + j\left(\omega((2.31\mathrm{E}-8)) - \omega^3(5.06\mathrm{E}-28)\right)}{1 - \omega^2(5.06\mathrm{E}-18) + j\omega(4.40\mathrm{E}-10)}$$

$$V_2 = \frac{V'R}{sL_2 + R}$$

$$Z_{12} = \left.\frac{V_2}{I_1}\right|_{I_2=0} = \frac{\dfrac{R}{sL_2 + R}}{\dfrac{1}{sL_2 + R} + sC} = \frac{R}{1 + sC(sL_2 + R)} = \frac{R}{1 + s^2CL_2 + sR}$$

$$Z_{12} = \frac{2}{1 - \omega^2(5.06\mathrm{E}-18) + j\omega 2} \ \Omega$$

$$I_2 = I_R + I_c = \frac{V_2}{R} + \frac{V_2}{sL_2 + \dfrac{1}{sC}} = \frac{V_2}{R} + \frac{sCV_2}{s^2L_2C + 1} = \frac{V_2\left(s^2L_2C + 1 + sRC\right)}{R\left(s^2L_2C + 1\right)}$$

$$Z_{22} = \left.\frac{V_2}{I_2}\right|_{I_1=0} = \frac{R\left(s^2L_2C + 1\right)}{s^2L_2C + 1 + sCR} = \frac{2\left(-\omega^2(2.3\mathrm{E}-8)(2.20\mathrm{E}-10) + 1\right)}{1 - \omega^2(2.3\mathrm{E}-8)(2.20\mathrm{E}-10) + j\omega 2(2.20\mathrm{E}-10)} \ \Omega$$

$$Z_{22} = \frac{\left(2 - \omega^2(1.012\mathrm{E}-17)\right)}{1 - \omega^2(5.06\mathrm{E}-18) + j\omega(4.40\mathrm{E}-10)} \ \Omega$$

$$Z_{21} = \left.\frac{V_1}{I_2}\right|_{I_1=0} = \frac{I_c\dfrac{1}{sC}}{I_2} = \frac{\left(\dfrac{V_2sC}{s^2L_2C + 1}\right)sC}{\dfrac{1}{R} + \dfrac{sC}{s^2L_2C + 1}} = \frac{s^2RC^2}{s^2L_2C + 1 + sRC}$$

$$Z_{21} = \frac{-2\omega^2(2.20\mathrm{E}-10)^2}{1 - \omega^2(2.3\mathrm{E}-8)(2.20\mathrm{E}-10) + j\omega 2(2.20\mathrm{E}-10)} \ \Omega$$

$$= \frac{-\omega^2(9.68\mathrm{E}-20)^2}{1 - \omega^2(5.06\mathrm{E}-18) + j\omega(4.40\mathrm{E}-10)} \ \Omega$$

2.3

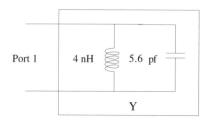

$$Y = \frac{I_1}{V_1} = sC + \frac{1}{sL} = \frac{1+s^2CL}{sL} = \frac{1-\omega^2(5.6E-12)(4E-9)}{j\omega(4E-9)} \text{ mho}$$

$$= \frac{j\left(\omega^2(22.4E-21)-1\right)}{\omega(4E-9)} \text{ mho}$$

2.4

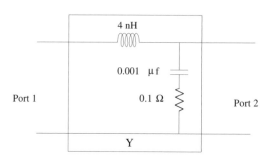

$$Y_{11} = \frac{I_1}{V_1}\bigg|_{V_2=0} = \frac{1}{sL} = \frac{-j}{\omega(4E-9)}$$

$$Y_{12} = \frac{I_2}{V_1}\bigg|_{V_2=0} = \frac{1}{sL} = \frac{-j}{\omega(4E-9)}$$

$$I_2 = \frac{V_2}{R+\dfrac{1}{sC}} + \frac{V_2}{sL} = \frac{V_2sC}{sRC+1+s^2LC}$$

$$Y_{22} = \frac{I_2}{V_2}\bigg|_{V_1=0} = \frac{sC}{sCR+1+s^2LC} = \frac{j\omega(1.0\text{E}-9)}{1-\omega^2(4.0\text{E}-18)+j\omega0.1(1.0\text{E}-9)}\text{ mho}$$

$$Y_{21} = \frac{I_1}{V_2}\bigg|_{V_1=0} = \frac{1}{sL} = \frac{-j}{\omega(4\text{E}-9)}$$

2.5

$$Z = \begin{bmatrix} \dfrac{sC(sL+R)+1}{sC} & \dfrac{s^2CL+1}{sC} \\ \dfrac{s^2CL+1}{sC} & \dfrac{s^2CL+1}{sC} \end{bmatrix}$$

$$\begin{bmatrix} \dfrac{sC(sL+R)+1}{sC} & \dfrac{s^2CL+1}{sC} \\ \dfrac{s^2CL+1}{sC} & \dfrac{s^2CL+1}{sC} \end{bmatrix}\begin{vmatrix} 1 & 0 \\ 0 & 1 \end{vmatrix}$$

$$\begin{bmatrix} sC(sL+R)+1 & s^2CL+1 \\ s^2CL+1 & s^2CL+1 \end{bmatrix}\begin{vmatrix} sC & 0 \\ 0 & sC \end{vmatrix}$$

$$\begin{bmatrix} sCR & 0 \\ s^2CL+1 & s^2CL+1 \end{bmatrix}\begin{vmatrix} sC & -sC \\ 0 & sC \end{vmatrix}$$

$$\begin{bmatrix} 1 & 0 \\ 1 & 1 \end{bmatrix}\begin{vmatrix} \dfrac{1}{R} & \dfrac{-1}{R} \\ 0 & \dfrac{sC}{s^2CL+1} \end{vmatrix}$$

$$\begin{bmatrix} 1 & 0 \\ 0 & 1 \end{bmatrix}\begin{vmatrix} \dfrac{1}{R} & \dfrac{-1}{R} \\ \dfrac{-1}{R} & \dfrac{sC}{s^2CL+1}+\dfrac{1}{R} \end{vmatrix}$$

$$Y = \begin{bmatrix} \dfrac{1}{R} & \dfrac{-1}{R} \\ \dfrac{-1}{R} & \dfrac{sC}{s^2CL+1} + \dfrac{1}{R} \end{bmatrix} = \begin{bmatrix} \dfrac{1}{R} & \dfrac{-1}{R} \\ \dfrac{-1}{R} & \dfrac{sCR+1+s^2CL}{R(s^2CL+1)} \end{bmatrix}$$

2.6

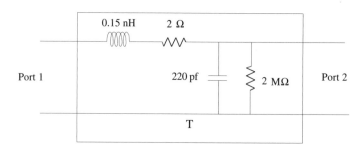

$$V_2 = V_1 \dfrac{R_2 + \dfrac{1}{sC}}{R_2 + \dfrac{1}{sC} + sL + R_1} = V_1 \dfrac{1+sCR_2}{1+sC(R_1+R_2)+s^2LC}$$

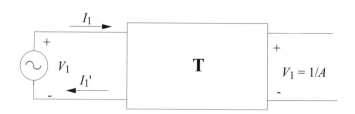

$$\frac{1}{A} = \frac{V_2}{V_1} = \frac{1+j\omega CR_2}{1-\omega^2 LC + j\omega C(R_1+R_2)} = \frac{1+j\omega(4.4\text{E}-4)}{1-\omega^2(3.30\text{E}-20)+j\omega(4.4\text{E}-4)}$$

$$A = \frac{1-\omega^2(3.30\text{E}-20)+j\omega(4.4\text{E}-4)}{1+j\omega(4.4\text{E}-4)}$$

$$I_2 = \frac{-V_1}{R+sL}$$

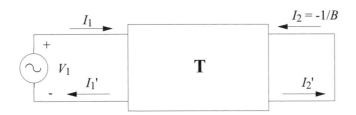

$$-\frac{1}{B} = \frac{I_2}{V_1} = \frac{-1}{R+sL}$$

$$B = R + sL = 2 + j\omega(1.5\text{E}-10)$$

$$V_2 = -I_1\left(R_2 + \frac{1}{sC}\right) = -I_1\left(\frac{R_2+1}{sC}\right) = I_1\left(\frac{-(R_2+1)}{j\omega(2.2\text{E}-10)}\right)$$

$$-\frac{1}{C} = \frac{V_2}{I_1} = \left(\frac{-3}{j\omega(2.2\text{E}-10)}\right)$$

$$C = j\omega(7.333\text{E}-11)$$

$$\frac{1}{D} = \frac{I_2}{I_1} = \frac{-I_1}{I_1}$$

$$D = 1$$

Solutions

3.1

$$V = 0.34 \exp\left(j3e7t - x(j35 + .3) \right)$$

$$\omega = 3E7$$

$$\beta = 35$$

$$\alpha = 0.3$$

$$v = \frac{3E7}{35} = 8.57E5 \text{ m/s}$$

$$\lambda = \frac{2\pi}{35} = 0.1795 \text{ m}$$

3.2

$$V = V_o \exp(-\gamma + j\omega t)$$

$$I = \frac{V_o}{300} \exp(-\gamma + j\omega t)$$

$$\beta = \frac{\omega}{v} = \frac{\omega}{1.0E8}$$

$$\gamma = x\left(j\frac{\omega}{1.0E8} + 10^{0.01}\right)$$

3.3

$$\gamma = \sqrt{(sL+R)(sC+G)} = \sqrt{(j\omega L+R)(j\omega C+G)}$$

$$= \sqrt{-\omega^2 LC + j\omega LG + j\omega RC + RG}$$

$$\gamma = \sqrt{(RG-\omega^2 LC) + j\omega(LG+RC)}$$

$$\text{using} \quad \sqrt{C+jD} = \left(\sqrt{\frac{C+D}{2}} + \sqrt{\frac{C-D}{2}}\right) + j\left(\sqrt{\frac{C+D}{2}} - \sqrt{\frac{C-D}{2}}\right)$$

$$\gamma = \left(\sqrt{\frac{(RG-\omega^2 LC)+\omega(LG+RC)}{2}} + \sqrt{\frac{(RG-\omega^2 LC)-\omega(LG+RC)}{2}}\right.$$

$$\left. +j\left(\sqrt{\frac{(RG-\omega^2 LC)+\omega(LG+RC)}{2}} - \sqrt{\frac{(RG-\omega^2 LC)-\omega(LG+RC)}{2}}\right)\right.$$

$$\chi = \left(\sqrt{\frac{(RG-\omega^2 LC)+\omega(LG+RC)}{2}} + \sqrt{\frac{(RG-\omega^2 LC)-\omega(LG+RC)}{2}}\right)$$

$$\beta = \left(\sqrt{\frac{(RG-\omega^2 LC)+\omega(LG+RC)}{2}} - \sqrt{\frac{(RG-\omega^2 LC)-\omega(LG+RC)}{2}}\right)$$

3.4

$$\gamma = j\beta + \alpha = j\frac{\omega}{1.2E8} + 10^{-0.025}$$

$$\gamma = \sqrt{(sL+R)(sC+G)} = (sC+G)Z_o = 75(j\omega C + G)$$

$$\beta = \frac{\omega}{1.2E8} = \omega C(75)$$

$$C = \frac{1}{(1.2E8)(75)} = 1.11E-10F$$

$$\alpha = 10^{-0.025} = G(75)$$

$$G = \frac{10^{-0.025}}{75} = 0.0126\text{mho}$$

$$Z_o^2 = (75)^2 = \frac{sL+R}{sC+G}$$

$$j\omega L + R = (j\omega C + G)(75^2)$$

$$R = G(75)^2 = 70.8\Omega$$

$$L = C(75)^2 = 6.25E-7H$$

3.5

$$\Gamma = \frac{75-50}{75+50} = 0.20$$

$$\text{VSWR} = \frac{|a|+|b|}{|a|-|b|} = \frac{|\Gamma||b|+|b|}{|\Gamma||b|-|b|} = \frac{|\Gamma|+1}{|\Gamma|-1} = \frac{0.2+1}{0.2-1} = 1.50$$

3.6

$$Z_L = \frac{R}{sRC+1}$$

$$\Gamma = \frac{\dfrac{R}{sCR+1}-50}{\dfrac{R}{sCR+1}+50} = \frac{R-50(sCR+1)}{R+(sCR+1)} = \frac{50-50(j\omega(1.00E-10)(50)+1)}{50+50(j\omega(1.00E-10)(50)+1)} :$$

$$= \frac{-j\omega(5.0\text{E}-9)}{2+j\omega(5.0\text{E}-9)}$$

3.7

$$\Gamma_o = \frac{300-75}{300+75} = 0.60$$

$$\Gamma_{(x)} = \Gamma_o\left(\cos 2\beta x - j\sin 2\beta x\right) = 0.6\left(\cos 2\beta x - j\sin 2\beta x\right)$$

3.8

A) $10 - j70$
B) $80 + j10$
C) $40 + j40$
D) $20 - j80$
E) $30 - j0$

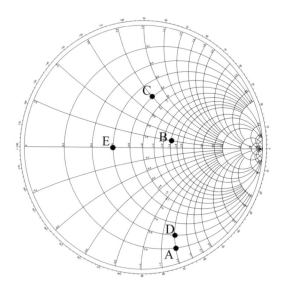

3.9

$$\Gamma_{in} = \frac{75-50}{75+50} = 0.20$$

3.10

Constant 30Ω

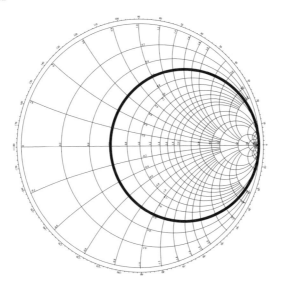

3.11

$$Y = \begin{bmatrix} \dfrac{1}{z} & -\dfrac{1}{z} \\[2ex] -\dfrac{1}{z} & \dfrac{1}{z} \end{bmatrix} \qquad \text{where} \qquad z = \frac{Z}{Z_o}$$

$$S_{11} = S_{22} = \frac{\left(1-\dfrac{1}{z}\right)\left(1+\dfrac{1}{z}\right)+\left(\dfrac{1}{z}\right)^2}{\left(1+\dfrac{1}{z}\right)\left(1+\dfrac{1}{z}\right)-\left(\dfrac{1}{z}\right)^2} = \frac{(z-1)(z+1)+1}{(z+1)(z+1)-1} = \frac{z^2}{z(z+2)} = \frac{z}{z+2} = \frac{Z}{Z+2Z_o}$$

$$S_{12} = S_{21} = \frac{2\dfrac{1}{z}}{\left(1+\dfrac{1}{z}\right)\left(1+\dfrac{1}{z}\right)-\left(\dfrac{1}{z}\right)^2} = \frac{2}{(z+1)(z+1)-1} = \frac{2Z_o}{Z+2Z_o}$$

3.12

$$Z = \begin{bmatrix} z & z \\ z & z \end{bmatrix} \quad \text{where} \quad z = \frac{Z}{Z_o}$$

$$S_{11} = S_{22} = \frac{(z-1)(z+1)-z^2}{(z+1)(z+1)-z^2} = \frac{-1}{2z+1} = \frac{-Z_o}{2Z+Z_o}$$

$$S_{12} = S_{21} = \frac{2z}{(z+1)(z+1)-z^2} = \frac{2z}{2z+1} = \frac{2Z}{2Z+Z_o}$$

3.13

$$Z_{in} = 40 \frac{60 + j40 \tan\left(\frac{2\pi}{\lambda}\frac{\lambda}{4}\right)}{40 + j60 \tan\left(\frac{2\pi}{\lambda}\frac{\lambda}{4}\right)} \quad \text{as} \quad \tan\left(\frac{2\pi}{\lambda}\frac{\lambda}{4}\right) \to \infty \quad \frac{(40)^2}{60} = 26.67$$

$$\Gamma = \frac{26.667 - 50}{26.667 + 50} = -0.3043$$

3.14

$$44 - 2(9.9) = 24.2$$

$$H' = \frac{Z\sqrt{2(9.9+1)}}{119.9} - \frac{1}{2}\left(\frac{9.9-1}{9.9+1}\right)\left(\ln\frac{2}{\pi} + \frac{1}{9.9}\ln\frac{4}{\pi}\right) = Z(0.0389) + 0.1744$$

$$Z_o = 50, \quad H' = 50(0.0389) + 0.1744 = 2.1215$$

$$\frac{w}{h} = \left(\frac{\exp(1.3426)}{8} - \frac{1}{4\exp(1.3426)}\right)^{-1} = 2.4193$$

$Z_0 = 50, \quad H' = 50(0.0389) + 0.1744 = 2.1215$

$$\frac{w}{h} = \left(\frac{\exp(2.1215)}{8} - \frac{1}{4\exp(2.1215)} \right)^{-1} = 0.9872$$

$Z_0 = 75, \quad H' = 75(0.0389) + 0.1744 = 3.095$

$$\frac{w}{h} = \left(\frac{\exp(3.095)}{8} - \frac{1}{4\exp(3.095)} \right)^{-1} = 0.3637$$

$Z_0 = 100, \quad H' = 100(0.0389) + 0.1744 = 4.0685$

$$\frac{w}{h} = \left(\frac{\exp(4.0685)}{8} - \frac{1}{4\exp(4.0685)} \right)^{-1} = 0.1369$$

3.15

$$\frac{h}{w} = \frac{0.6}{1.4} = .4286$$

$$Z_0 = \frac{119.9}{\sqrt{2(2.2+1)}} \left\{ \ln\left[4(0.4286) + \sqrt{16(0.4286)^2 + 2} \right] \right.$$

$$\left. - \frac{1}{2}\left(\frac{2.2-1}{2.2+1} \right)\left(\ln\frac{\pi}{2} + \frac{1}{2.2}\ln\frac{4}{\pi} \right) \right\} = 60.0\,\Omega$$

Solutions

4.1

$x = 3$ on the w plane $\quad w = \dfrac{1}{s}$

$$v3 = -uy$$

$$y = -\frac{3v}{u}$$

$$u3 - \frac{v^2 3}{u} = 1$$

$$u^2 3 - v^2 3 = u$$

$$u^2 - \frac{u}{3} - v^2 = 0$$

$$u^2 - \frac{u}{3} + \frac{1}{36} - v^2 = \frac{1}{36}$$

$$\left(u - \frac{1}{6}\right)^2 - v^2 = \left(\frac{1}{6}\right)^2 \quad \text{is a circle with the center at} \quad \left(\frac{1}{6}, 0\right)$$

and a radius of $\dfrac{1}{6}$

4.2

$y = 5$ on the w-plane $\qquad w = \dfrac{2 + j41}{s}$

$s = x + j5$

$\dfrac{2 + j4}{x + j5} = u + jv$

$2 + j4 = xu + jvx + j5u - 5v$

Equate the real and imaginary parts.

$2 = xu - 5v$

$x = \dfrac{2 + 5v}{u}$

$4 = vx + 5u$

$4 = v\left(\dfrac{2 + 5v}{u}\right) + 5u$

$2v + 5v^2 + 5u^2 - 4u = 0$

$u^2 - \dfrac{4}{5}u + v^2 + \dfrac{2}{5}v = 0$

$u^2 - \dfrac{4}{5}u + \dfrac{16}{100} + v^2 + \dfrac{2}{5}v + \dfrac{4}{100} = \dfrac{2}{10}$

$\left(u - \dfrac{4}{10}\right)^2 + \left(v + \dfrac{2}{10}\right)^2 = \dfrac{2}{10}$ which is a circle with a center at $\left(\dfrac{4}{10}, \dfrac{2}{10}\right)$

and a radius of $\sqrt{\dfrac{2}{10}}$

4.3

$3x + 2 = y$ mapped on the w-plane with $w = \dfrac{1}{3 + j4 + s}$

$s = x + jy = x + j(3x + 2)$

$u + jv = \dfrac{1}{2 + j4 + x + j(3x + 2)} = \dfrac{1}{(3 + x) + j(3x + 6)}$

$(u + jv)\big((3 + x) + j(3x + 6)\big) = 1$

$u(3 + x) + ju(3x + 6) + jv(3 + x) - v(3x + 6) = 1$

Equate the real and imaginary parts.

$3u + xu - 3xv - 6v = 1$

$3xu + 6u + 3v + xv = 0$

$x = \dfrac{-6u - 3v}{3u + v}$

$3u - u\dfrac{6u + 3v}{3u + v} + 3v\dfrac{6u + 3v}{3u + v} - 6v = 1$

$3u(3u + v) - 6u^2 - 3uv + 3v(6u + 3v) - 6v(3u + v) = 1$

$3u^2 + 3v^2 = 1$

$u^2 + v^2 = \dfrac{1}{3}$ which is a circle with a center at (0,0) and a radius of $\sqrt{\dfrac{1}{3}}$

4.4

Center of $6 + j7$, $R = 2$ and $w = \dfrac{1}{s}$

$a = 0$, $b = 1$, $c = 0$, $d = 0$, $0 = 6 + j7$

$Q = \dfrac{|2|^2(0)(0 + 1(6 + j7)) - 1(1 + 0(6 + j7))^*}{|2|^2|1|^2 - |(1)(6 + j7) + 0|^2} = \dfrac{-6 + j7}{4 - (36 + 49)} = \dfrac{6}{71} + j\dfrac{7}{71}$

$$Q = \frac{\left|2\right|^2 \left|(0)(0 + 1(6 + j7)) - 1(1 + 0(6 + j7))\right|}{\left|2\right|^2 \left|1\right|^2 - \left|(1)(6 + j7) + 0\right|^2} = \frac{4}{71}$$

4.5

Center at $3 - j2$, $R = 3$ and $w = \dfrac{s}{2s + j6 - 3}$

a = 1, b = 0, c =1, d = $-3 + j6$

$$Q = \frac{\left|3\right|^2 (1)\left(1^*\right) - \left(0 + 1(3 - j2)(1(3 - j2) + (-3 + j6))\right)^*}{\left|3\right|^2 \left|1\right|^2 - \left|(1)(3 - j2) + (-3 + j6)\right|^2}$$

$$Q = \frac{9 - (3 - j2)(j4))^*}{9 - \left|j4\right|^2} = \frac{9 + 8 + j12}{-7} = \frac{-17 - j12}{7}$$

$$R = \frac{\left|3\right|^2 \left|(1)((-3 + j6) + 1(3 - j2)) - 1(0 + 1(3 + j2))\right|}{\left|3\right|^2 \left|1\right|^2 - \left|(1)(3 - j2) + (-3 + j6)\right|^2}$$

$$R = \frac{\left|3\right|^2 \left|-3 + j6 + 3 - j2 - 3 - j2\right|}{7} = \frac{9\left|-3 + j6\right|}{7} = \frac{9}{7}\sqrt{9 + 36} = \frac{9\sqrt{45}}{7}$$

4.6

Two circles centered at the origin with a radius of $R = 0.25$ and $R = 0.15$ with

$$w = \frac{2 + j4}{s}$$

First circle with a raduis of $R = 0.15$:

$$Q = \frac{\left|0.15\right|^2 (0)(1^*) - (0)(2 + j4)}{\left|0.15\right|^2 \left|1\right|^2 - \left|0\right|^2} = 0$$

$$R = \frac{\left|0.15\right| \left|(0)(0) - (2 + j4)(1)\right|}{\left|0.15\right|^2 \left|1\right|^2 - \left|0\right|^2} = \frac{\left|2 + j4\right|}{0.15} = \frac{\sqrt{20}}{0.15}$$

Second circle R = 0.25:

$$Q = \frac{\left|0.25\right|^2 (0)(1^*) - (0)(2+j4)}{\left|0.25\right|^2 \left|1\right|^2 - \left|0\right|^2} = 0$$

$$R = \frac{\left|0.25\right| \left|(0)(0) - (2+j4)(1)\right|}{\left|0.25\right|^2 \left|1\right|^2 - \left|0\right|^2} = \frac{\left|2+j4\right|}{0.25} = \frac{\sqrt{20}}{0.25}$$

4.7

Three measurements: load (M_1), short (M_2) and open (M_3)

Using Figure 4-12(f): $M_x = e_{11} + \dfrac{e_{12}e_{21}\Gamma_L}{1 - e_{22}\Gamma_L}$

When $\Gamma_L = 0$, $M_1 = e_{11}$
When $\Gamma_L = 1$ (open)

$$M_3 = M_1 + \frac{e_{12}e_{21}\Gamma_L}{1 - e_{22}\Gamma_L}$$

$$\left(M_2 - M_1\right)\left(1 - e_{22}\right) = e_{12}e_{21}$$

When $\Gamma_L = -1$ (open)

$$M_3 = M_1 + \frac{e_{12}e_{21}\Gamma_L}{1 - e_{22}\Gamma_L}$$

$$\left(M_3 - M_1\right)\left(1 + e_{22}\right) = -e_{12}e_{21}$$

$$\left(M_2 - M_1\right) - e_{22}\left(M_2 - M_1\right) = \left(M_1 - M_3\right) - e_{22}\left(M_3 - M_1\right)$$

$$e_{22} = \frac{\left(M_2 - M_1\right) - \left(M_1 - M_3\right)}{\left(M_2 - M_1\right) - \left(M_3 - M_1\right)} = \frac{M_2 + M_3 - 2M_1}{M_2 - M_3}$$

$$e_{12}e_{21} = \left(M_2 - M_1\right)\left(1 - \frac{M_2 + M_3 - 2M_1}{M_2 - M_3}\right) = \left(M_2 - M_1\right)\left(\frac{2M_1 - 2M_3}{M_2 - M_3}\right)$$

$$e_{12} = e_{21} = \sqrt{(M_2 - M_1)\left(\frac{2M_1 - 2M_3}{M_2 - M_3}\right)}$$

4.8

$$\Gamma = \frac{Z_L - Z_o}{Z_L - Z_o} = \frac{1 - \dfrac{Z_o}{Z_L}}{1 + \dfrac{Z_o}{Z_L}} = \frac{\dfrac{1}{Z_o} - \dfrac{1}{Z_L}}{\dfrac{1}{Z_o} + \dfrac{1}{Z_L}} = \frac{Y_o - Y_L}{Y_o + Y_L}$$

$$S_{11} = S_{22} = \Gamma = \frac{Y_o - Y - Y_o}{Y_o + Y + Y_o} = \frac{-Y}{2Y_o + Y} = \frac{-y}{2 + y}$$

$$I_L = I_s - I_1$$

$$V_L = V_s$$

$$a_1 = \frac{1}{2}\left(\frac{V_1 + Z_o I_1}{\sqrt{Z_o}}\right)$$

$$a_1 = \frac{1}{2}\left(\frac{V + Z_o(VY + VY_0)}{\sqrt{Z_o}}\right)$$

$$b_2 = \frac{1}{2}\left(\frac{V_2 - Z_o(-VY_0)}{\sqrt{Z_o}}\right)$$

$$S_{12} = S_{21} = \frac{b_2}{a_1} = \frac{V + Z_o(VY_o)}{V + Z_o(VY + VY_o)} = \frac{1 + Z_o Y_o}{1 + Z_o Y + 1} = \frac{2}{2 + y}$$

4.9

$$S_{11} = S_{22} = \Gamma = \frac{Z_L - Z_o}{Z_L - Z_o} = \frac{Z + Z_o - Z_o}{Z + Z_o + Z_o} = \frac{Z}{Z + 2Z_o} = \frac{z}{z + 2}$$

$$a_1 = \frac{1}{2}\left(\frac{(IZ_o + IZ) + Z_o I}{\sqrt{Z_o}}\right)$$

$$b_2 = \frac{1}{2}\left(\frac{IZ_o - Z_o(-I)}{\sqrt{Z_o}} \right)$$

$$S_{12} = S_{21} = \frac{b_2}{a_1} = \frac{2Z_o I}{IZ + 2Z_o I} = \frac{2Z_o}{Z + 2Z_o} = \frac{2}{2+z}$$

Solutions

5.1

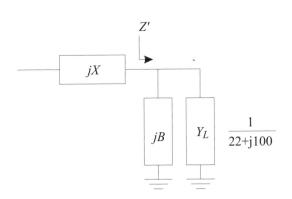

$Y = 0.0021 - j0.0095$

$B = 0.0095 + \sqrt{0.02(0.0021) - (0.0021)^2} = 0.0157\text{mho}$

$X = -\dfrac{50(0.0157 - 0.0095)}{0.0021} = -146.4\Omega$

or

$B = 0.0095 - \sqrt{0.02(0.0021) - (0.0021)^2} = 0.0034\text{mho}$

$X = -\dfrac{50(0.0034 - 0.0095)}{0.0021} = 146.0\Omega$

5.2

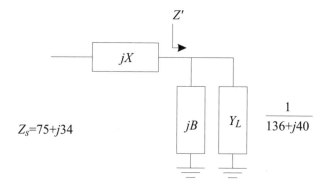

Re{Z}' needs to be 75Ω.
$Y_L = 0.0068 - j0.0020$

$$B = -0.0020 - \sqrt{\frac{1}{75}(0.0068) - (0.0068)^2} = 0.0047\text{mho}$$

$$= -40 - \frac{75(0.0047 + 0.0020)}{0.0068} = 107.87\Omega$$

or

$$B = -0.0020 + \sqrt{\frac{1}{75}(0.0068) - (0.0068)^2} = -0.0087\text{mho}$$

$$X = -40 - \frac{75(-0.0087 + 0.0020)}{0.0068} = -39.87\Omega$$

5.3

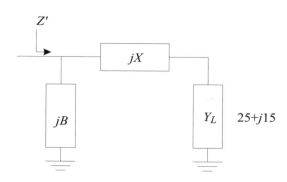

$$X = -15 + \sqrt{100(25) - (25)^2} = 28.3\Omega$$

$$B = \frac{\dfrac{1}{100}(28.3 + 15)}{25} = 0.0173\text{mho}$$

or

$$X = -15 - \sqrt{100(25) - (25)^2} = -51.3\Omega$$

$$B = \frac{\dfrac{1}{100}(-51.3 + 15)}{25} = -0.0173\text{mho}$$

5.4

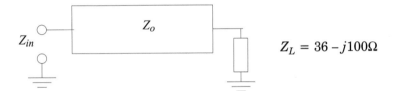

$$Z_t = \sqrt{75(36) - \frac{(0)^2 36 + (110)^2 75}{75 - 110}} = 169.2\Omega$$

$$\tan \beta x = \frac{169.2 * (75 - 36)}{75 * 110 - 36 * (0)} = 0.799$$

$$\beta x = 0.6747 = \frac{2\pi}{\lambda} x$$

$$x = 0.1074\lambda$$

5.5

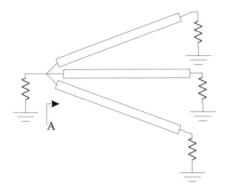

$$Z_o = \sqrt{(150)(50)} = 86.60$$

5.6

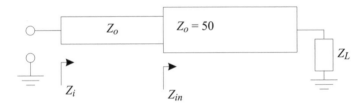

$$Z_o = \frac{Z_t^2}{Z_L}$$

$$50 = \frac{Z_L^2(50 + jZ_L \tan \beta x)}{50Z_L + j(50)^2 \tan \beta x}$$

$$\frac{Z_{in}}{Z_o} = \frac{G_L + j(B_L + 50 \tan \beta x)}{(50 - B_L \tan \beta x) + jG_L \tan \beta x}$$

$$\frac{Z_{in}}{Z_o} = \frac{(G_L + j(B_L + 50 \tan \beta x))((50 - B_L \tan \beta x) + jG_L \tan \beta x)}{(50 - B_L \tan \beta x)^2 + (G_L \tan \beta x)^2}$$

Equate the real and imaginary parts.

$$0 = (B_L + 50 \tan \beta x)(50 - B_L \tan \beta x) + G_L^2 \tan \beta x$$

$$0 = 50B_L + \left(50^2 + G_L^2 - B_L^2\right)\tan\beta x - 50B_L \tan^2\beta x$$

$$\tan\beta x = \frac{-\left(50^2 + G_L^2 - B_L^2\right) \pm \sqrt{\left(50^2 + G_L^2 - B_L^2\right)^2 + 4\left(50B_L\right)^2}}{2\left(50B_L\right)}$$

$$Z_{in} = \frac{Z_o\left[G_L\left(50 - B_L \tan\beta x\right) - G_L \tan\beta x\left(B_L + 50\tan\beta x\right)\right]}{\left(50 - B_L \tan\beta x\right)^2 - \left(G_L \tan\beta x\right)^2}$$

$$Z_t = \sqrt{Z_o Z_{in}} = \sqrt{\frac{50G_L\left(1 - \tan\beta x\right) - 2B_L G_L \tan\beta x}{\left(50 - B_L \tan\beta x\right)^2 - \left(G_L \tan\beta x\right)^2}}$$

5.7

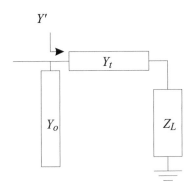

$$Y_t = \frac{1}{75}$$

$$\tan\beta x = \frac{-35 + \sqrt{\dfrac{25}{75}\left[\left(75 - 25\right)^2 + \left(35\right)^2\right]}}{25 - 75} = -0.0047$$

$$\beta x = -0.0047 = \frac{2\pi}{\lambda}x$$

$$x = -7.4802\mathrm{E} - 4\lambda \quad \text{i.e.} \quad 0.4993\lambda$$

$$\text{Im}\{Y'\} = \frac{(25^2)(-0.0047) - (75 + 35(-0.0047))(-35 + 75(-0.0047))}{75[(25)^2 + (-35 + 75(-0.0047))^2]} = 0.0188$$

$$-0.0188 = \frac{1}{75}\tan\frac{2\pi x}{\lambda}$$

$$x = -0.152\lambda \quad \text{i.e.} \quad 0.343\lambda$$

or

$$\tan\beta x = \frac{-35 - \sqrt{\dfrac{25}{75}\left[(75-25)^2 + (35)^2\right]}}{25 - 75} = 1.4047$$

$$\beta x = 0.9521 = \frac{2\pi}{\lambda}x$$

$$x = 0.152\lambda$$

$$\text{Im}\{Y'\} = \frac{(25^2)(1.4047) - (75 + 35(1.4047))(-35 + 75(1.4047))}{75[(25)^2 + (-35 + 75(1.4047))^2]} = -0.0188$$

$$0.0188 = \frac{1}{75}\tan\frac{2\pi x}{\lambda}$$

$$x = 0.152\lambda$$

5.8

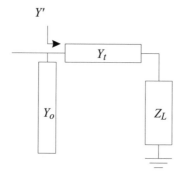

$$T_s = G_s + jB_s$$

$$Z_L = R_L + jX_L$$

$$\text{Re}\{Y'\} = G_s = Y_o \frac{Z_o R_L \left(1 + \tan^2 \beta x\right)}{R_L^2 + \left(X_L + Z_o \tan \beta x\right)^2}$$

$$G_s \left[R_L^2 + X_L^2 + 2X_L Z_o \tan \beta x + Z_o^2 \tan^2 \beta x \right] = R_L + R_L \tan^2 \beta x$$

$$0 = \left[G_s Z_o^2 - R_L \right] \tan^2 \beta x + 2G_s X_L Z_o \tan \beta x + G_s R_L^2 + G_s X_L^2 - R_L$$

$$\tan \beta x = \frac{-2G_s X_L Z_o \pm \sqrt{\left(2G_s X_L Z_o\right)^2 - 4\left(G_s Z_o^2 - R_L\right)\left(G_s R_L^2 + G_s X_L^2 - R_L\right)}}{2\left(G_s Z_o^2 - R_L\right)}$$

$$\tan \beta x = \frac{X_L Z_o \pm \sqrt{\dfrac{R_L}{G_s}\left(Z_o^2 + R_L^2 + X_L^2 - \dfrac{R_L}{G_s}\right) - Z_o^2 R_L^2}}{\left(\dfrac{R_L}{G_s} - Z_o^2\right)}$$

$$\text{Im}\{Y'\} = Y_o \frac{R_L^2 - \left(Z_o - X_L \tan \beta x\right)\left(X_L + Z_o \tan \beta x\right)}{R_L^2 + \left(X_L + Z_o \tan \beta x\right)^2}$$

$$Y_o \tan \beta x' = Y_o \frac{\left(Z_o - X_L \tan \beta x\right)\left(X_L + Z_o \tan \beta x\right) - R_L^2}{R_L^2 + \left(X_L + Z_o \tan \beta x\right)^2}$$

$$x' = \frac{\lambda}{2\pi} \tan^{-1} \frac{\left(Z_o - X_L \tan \beta x\right)\left(X_L + Z_o \tan \beta x\right) - R_L^2}{R_L^2 + \left(X_L + Z_o \tan \beta x\right)^2}$$

5.9

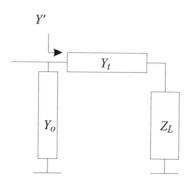

$$\tan \beta x = \frac{32 + \sqrt{\dfrac{115}{75}\left[(75-115)^2 + (32)^2\right]}}{115-75} = 2.3858$$

$$x = \frac{\lambda}{2\pi} \tan^{-1}(2.3858) = 0.1868\lambda$$

$$\text{Im}\{Y'\} = \frac{(115)^2(2.3858) - (50 - 32(2.3858))(32 + 50(2.3858))}{50\left(115^2 + (32 + 50(2.3858))^2\right)} = 0.0079$$

$$-0.0079 = \frac{1}{75} \cot \frac{2\pi x'}{\lambda}$$

$$x' = 0.170\lambda$$

or

$$\tan \beta x = \frac{32 - \sqrt{\dfrac{115}{75}\left[(75-115)^2 + (32)^2\right]}}{115-75} = -0.7858$$

$$x = \frac{\lambda}{2\pi} \tan^{-1}(-0.7858) = -0.1060\lambda \quad \text{i.e.} \quad 0.3940\lambda$$

$$\text{Im}\{Y'\} = \frac{(115)^2(-0.7858) - (50 - 32(-0.7858))(32 + 50(-0.7858))}{50\left(115^2 + (32 + 50(-0.7858))^2\right)} = -0.0074$$

$$x' = \frac{\lambda}{2\pi}\cot^{-1}(75*(-0.0074)) = -0.170\lambda \quad \text{i.e.} \quad 0.330\lambda$$

5.10

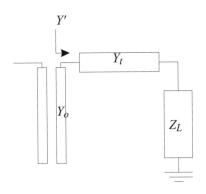

$$Z_s = R_s + jX_s$$

$$Y_L = G_L + jB_L$$

$$R_s = Z_o \frac{Y_o G_L\left(1 + \tan^2 \beta x\right)}{G_L^2 + \left(B_L + Y_o \tan \beta x\right)^2}$$

$$R_s\left(G_L^2 + \left(B_L + Y_o \tan \beta x\right)^2\right) = G_L\left(1 + \tan^2 \beta x\right)$$

$$0 = R_s G_L^2 + R_s B_L^2 - G_L + 2R_s B_L Y_o \tan \beta x + \left(Y_o R_s - G_L\right)\tan^2 \beta x$$

$$\tan \beta x = \frac{-2R_s B_L Y_o \pm \sqrt{\left(2R_s B_L Y_o\right)^2 - 4\left(R_s G_L^2 + R_s B_L^2 - G_L\right)\left(Y_o R_s - G_L\right)}}{2\left(Y_o R_s - G_L\right)}$$

$$\tan \beta x = \frac{R_s B_L Y_o \pm \sqrt{\left(R_s B_L Y_o\right)^2 - \left(R_s G_L^2 + R_s B_L^2 - G_L\right)\left(Y_o R_s - G_L\right)}}{\left(G_L - Y_o R_s\right)}$$

$$\text{Im}\{Y'\} = Z_o \frac{Y_o G_L\left(1 + \tan^2 \beta x\right)}{G_L^2 + \left(B_L + Y_o \tan \beta x\right)^2}$$

5.11

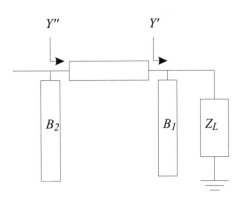

$$\tan\frac{2\pi}{\lambda}\frac{1}{4}\lambda = \tan\frac{3\pi}{4} = \infty$$

$$B_1\big|_{\tan\beta x\to\infty} = 16 + \sqrt{\frac{400}{50} - 50^2} = 66.08$$

$$66.08 = 50\cot\frac{2\pi x}{\lambda}$$

$$x'' = \frac{\lambda}{2\pi}\cot^{-1}\left(\frac{66.08}{50}\right) = 0.1031\lambda$$

$$B_2\big|_{\tan\beta x\to\infty} = \frac{B_L + B_1 - Y_t}{(B_L + B_1)^2 + G_L^2} = \frac{-16 + 66.08 - 50}{(-16 + 66.08)^2 + \left(\frac{1}{400}\right)^2} = 3.1872\mathrm{E}-005$$

$$3.1872\mathrm{E}-005 = 50\cot\frac{2\pi x}{\lambda}$$

$$x'' = \frac{\lambda}{2\pi}\cot^{-1}\left(\frac{3.1872\mathrm{E}-005}{50}\right) = 0.25\lambda$$

5.12

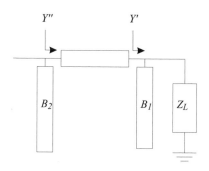

$$\tan\frac{2\pi}{\lambda}\frac{3}{8}\lambda = \tan\frac{3\pi}{4} = -1$$

$$\mathrm{Re}\{Y''\} = G_s$$

$$Y'' = \frac{1}{50}\frac{50 - j\big((32 - j(12 + B_1))\big)}{32 - j(12 + B_1) + j50} = \frac{1}{50}\frac{62 + B_1 - j32}{32 - j(62 + B_1)}$$

$$Y'' = \frac{(62 + B_1 - j32)(32 + j(62 + B_1))}{50\big(32^2 - (62 + B_1)^2\big)}$$

$$Y'' = \frac{32(62) + 32B_1 + 32(62 + B_1) + j\big(-32^2 + (62 + B_1)(62 + B_1)\big)}{50\big(32^2 - (62 + B_1)^2\big)}$$

$$\mathrm{Re}\{Y''\} = 95 = \frac{32(62) + 32B_1 + 32(62 + B_1)}{50\big(32^2 - (62 + B_1)^2\big)}$$

$$\mathrm{Re}\{Y''\} = 95(50)\big(32^2 - (62 + B_1)^2\big) = 32(64) + 32B_1 + 32(62 + B_1)$$

$$\mathrm{Re}\{Y''\} = \big(32^2 95(50) - 62^2 95(50) - 2(62)95(50)B_1 - 95(50)B_1^2\big) = 2(32)(64) + 64B_1$$

$$\mathrm{Im}\{Y''\} = \frac{-32^2 + (62 + B_1)(62 + B_1)}{50\big(32^2 - (62 + B_1)^2\big)}$$

6

Solutions

6.1

| 4.0 | 0.95 | -50 | 2.95 | 138 | 0.07 | 59 | 0.69 | -30 |

$\Delta = (0.95@-50°)(0.69@-30°) - (2.95@138°)(0.07@59°) = 0.6628@-62.0°$

$$C_s = \frac{(0.95@-50°)*-(\Delta)*(0.69@-30°)}{(0.95)^2-(\Delta)^2} = 1.1532@65.4°$$

$$R_s = \left|\frac{(2.95)(0.07)}{|0.95|^2-|\Delta|^2}\right| = 0.4458$$

6.2

| 6.0 | 0.89 | -70 | 2.67 | 120 | 0.09 | 47 | 0.60 | -42 |

$\Delta = (0.89@-70°)(0.60@-42°) - (2.67@120°)(0.09@47°) = 0.5502@-86.4°$

$$C_s = \frac{(0.89@-70°)*-(\Delta)*(0.60@-42°)}{(0.89)^2-(\Delta)^2} = 1.2446@83.5°$$

$$R_s = \left|\frac{(2.67)(0.09)}{|0.89|^2-|\Delta|^2}\right| = 0.4911$$

$$C_L = \frac{(0.60@-42°)*-(\Delta)*(0.89@-70°)}{|0.60|^2 - |\Delta|^2} = 4.6099@95.2°$$

$$R_L = \left| \frac{(2.67)(.09)}{|0.60|^2 - |\Delta|^2} \right| = 4.1976$$

6.3

$$\Delta = (0.99@-27°)(0.67@-16°) - (3.19@158°)(0.04@74°) = 0.6645@-32.0°$$

$$C_s = \frac{(0.99@-27°)*-(\Delta)*(0.67@-16°)}{(0.99)^2 - (\Delta)^2} = 1.0389@35.75°$$

2.0	0.99	-27	3.19	158	0.04	74	0.67	-16

$$R_s = \left| \frac{(3.19)(0.04)}{|0.99|^2 - |\Delta|^2} \right| = 0.2369$$

$$C_L = \frac{(0.67@-16°)*-(\Delta)*(0.99@-27°)}{|0.67|^2 - |\Delta|^2} = 17.310@95.05°$$

$$R_L = \left| \frac{(3.19)(.04)}{|0.67|^2 - |\Delta|^2} \right| = 17.233$$

6.4

$$S_{11} = .933@-108.3° \quad S_{21} = 1.928@72.3° \quad S_{12} = .046@15.5° \quad S_{22} = .724@-81.6°$$

$$\Delta = (0.933@-108.3°)(0.724@-81.6°) - (1.928@1 - 72.3°)(0.046@-15.5°)$$
$$= 0.6995@177.6.0°$$

$$K = \frac{1 - 0.933^2 - 0.724^2 + \Delta^2}{2(1.928)(0.046)} = 0.3013$$

6.5

$$S_{11} = .92@-69° \quad S_{21} = 2.34@112° \quad S_{12} = .047@43° \quad S_{22} = .63@-52°$$

$$\Delta = (0.92@-69°)(0.63@-52°) - (2.34@112°)(0.047@-43°) = 0.5785@-110.1°$$

$$K = \frac{1 - 0.92^2 - 0.63^2 + \Delta^2}{2(2.34)(0.047)} = .4156$$

$$C_s = \frac{(0.92@-69°)*-(\Delta)*(0.63@-52°)}{(0.92)^2 - (\Delta)^2} = 1.107@76.0°$$

$$R_s = \left| \frac{(2.34)(0.047)}{|0.92|^2 - |\Delta|^2} \right| = 0.2149$$

$$C_L = \frac{(0.63@-52°)*-(\Delta)*(0.92@-69°)}{|0.63|^2 - |\Delta|^2} = 2.3657@95.15°$$

$$R_L = \left| \frac{(2.34)(.047)}{|0.63|^2 - |\Delta|^2} \right| = 1.768$$

6.6

14.0	0.74	-135	1.93	63	0.13	12	0.55	-81

$$\Delta = (0.74@-135°)(0.55@-81°) - (1.93@63°)(0.13@12°) = 0.3942@-179.5°$$

$$K = \frac{1 - 0.74^2 - 0.55^2 + 0.3942^2}{2(1.93)(0.13)} = 0.6084$$

Conjugate match does not exist.

6.7

8.0	0.86	-87	2.45	104	0.11	36	0.58	-53

$$\Delta = (0.86@-87°)(0.58@-53°) - (2.45@104°)(0.11@36°) = 0.5242@-109.6°$$

$$K = \frac{1 - 0.86^2 - 0.58^2 + 0.5242^2}{2(2.45)(0.11)} = 0.3687$$

Conjugate match does not exist.
MSG = 22.27

6.8

For the transistor in Problem 6.5:

$$G_{T\max} = \frac{|2.34|}{|0.047|}\left(1.046 - \sqrt{1.0469^2 - 1}\right) = 41.34$$

For the transistor in Problem 6.6:

$$MSG = \frac{1.93}{0.13} = 14.85$$

6.9

$$\Gamma_s = \frac{75 - 50}{75 + 50} = 0.200@0.00°$$

$$\Gamma_L = 0.00@0.00°$$

$$\Delta = (0.11@\text{–}42°)(0.41@\text{–}31°) - (2.5@168°)(0.06@23°) = 0.1951@11.0°$$

$$\Gamma_{in} = \frac{\left(0.11@42°\right) - \left(0.195@11°\right)(0)}{\left(1 - (0.41@-31°)(0)\right)} = 0.11@42°$$

$$\Gamma_{out} = \frac{(0.41@-31°) - (0.195@11°)(0.2)}{\left(1 - (0.11@42°)(0.2)\right)} = 0.3882@-34.6°$$

$$G_t = \frac{1 - 0.2^2}{\left|1 - 0.11@42°(0.2)\right|^2}(2.5)^2\frac{1 - (0)^2}{\left|1 - (.41@31°)(0)\right|^2} = 6.1997$$

$$G_p = \frac{1}{1 - |0.11|^2}(2.5)^2\frac{1 - (0)^2}{\left|1 - (.41@31°)(0)\right|^2} = 6.3266$$

$$G_a = \frac{1 - (0.2)^2}{\left|1 - \left(0.11@42°\right)0.2\right|^2}(2.5)^2\frac{1}{1 - |0.3882|^2} = 7.2998$$

$$G_{TU} = \frac{1-(0.2)^2}{\left|1-(.11@42°)(0.2)\right|^2}(2.5)^2\frac{1-(0)^2}{\left|1-(0.41@-31°)(0)\right|^2} = 6.1197$$

$$K = \frac{1-0.11^2-0.41^2+(0.1841)^2}{2(2.5)(0.06)} = 3.0856$$

$$G_{T\max} = \frac{|S_{21}|}{|S_{12}|}\left(K-\sqrt{K^2-1}\right) = 6.9398$$

$$MSG = \frac{2.5}{0.06} = 41.67$$

6.10

$$\Gamma_s = \frac{40-j20-50}{40-j20+50} = 0.2425@-104.3°$$

$$\Gamma_L = \frac{70+j10-50}{70-j10+50} = 0.1857@-21.8°$$

$$\Delta = (0.11@-42°)(0.41@-31°) - (2.5@168°)(0.06@23°) = 0.1951@11.0°$$

$$\Gamma_{in} = \frac{(0.11@42°)-(0.195@11°)(0.1857@-21.8°)}{(1-(0.41@-31°)(0.1857@-21.8°))} = 0.0970@56.5°$$

$$\Gamma_{out} = \frac{(0.41@-31°)-(0.195@11°)(0.2425@-104.3°)}{\left(1-(0.11@42°)(0.2425@-104.3°)\right)} = 0.3949@-21.21°$$

$$G_t = \frac{1-0.2425^2}{\left|1-\left(0.0970@56.5°\right)\left(0.2425@-104.3°\right)\right|^2}(2.5)^2$$

$$\frac{1-(0.1857)^2}{\left|1-(.41@31°)(0.1857@-21.8°)\right|^2} = 6.4159$$

$$G_p = \frac{1}{1-|0.0970|^2}(2.5)^2 \frac{1-(0.1857)^2}{\left|1-(.41@31°)(0.1857@-21.8°)\right|^2} = 6.6669$$

$$G_a = \frac{1-(0.2425)^2}{\left|1-(0.11@42°)(0.2425@-104.3°)\right|^2}(2.5)^2 \frac{1}{1-0.3949^2} = 7.1427$$

$$G_{TU} = \frac{1-(0.2425)^2}{\left|1-(.11@42°)(0.2425@-104.3°)\right|^2}(2.5)$$

$$\frac{1-(0.1857)^2}{\left|1-(0.41@-31°)(0.1857@-21.8°)\right|^2} = 6.3706$$

$\Delta = (0.11@-42°)(0.41@-31°) - (2.5@168°)(0.06@23°) = 0.1841@11.0°$

$$K = \frac{1-0.11^2 - 0.41^2 + (0.1841)^2}{2(2.5)(0.06)} = 3.0856$$

$$G_{T\max} = \frac{2.5}{0.06}\left(3.0856 - \sqrt{3.0856^2 - 1}\right) = 6.9398$$

$$\text{MSG} = \frac{2.5}{0.06} = 41.67$$

6.11

10.0	0.81	-104	2.24	90	0.12	29	0.57	-63

$\Delta = (0.81@-104°)(0.57@-63°) - (2.24@90°)(0.12@29°) = 0.4658@-133.3°$

$$K = \frac{1-0.81^2 - 0.57^2 + (0.4658)^2}{2(2.24)(0.12)} = 0.4390$$

Gains: 10, 7.94, 6.31, 3.98

$$C_a = \frac{10\left((0.81@104) - (0.4658@133.3°)(0.57@-63)\right)}{2.24^2 + 10(0.81^2 - 0.4658^2)} = 0.6454@118.0°$$

$$R_a = \frac{2.24\sqrt{2.24^2 - 2(10)0.429(2.24)(0.12) + 10^2(0.12)^2}}{\left|2.24^2 + 10\left(0.81^2 - 0.4658^2\right)\right|} = 0.4819$$

$$C_p = \frac{10\left((0.81@104) - (0.4658@133.3)(0.57@-63)\right)}{2.24^2 + 10\left(.81^2 - 0.4658^2\right)} = 0.5424@102.3°$$

$$R_p = \frac{2.24\sqrt{2.24^2 - 2(10)(0.439)(2.24)(0.12) + 10^2(0.12)^2}}{\left|2.24^2 + 10\left(0.81^2 - .4658^2\right)\right|} = 0.7437$$

$$C_a = \frac{7.94\left((0.81@104) - \left(0.4658@133.3°\right)(0.57@-63)\right)}{2.24^2 + 7.94\left(0.81^2 - 0.4658^2\right)} = 0.5671@118.0°$$

$$R_a = \frac{2.24\sqrt{2.24^2 - 2(7.94)0.429(2.24)(0.12) + 7.94^2(0.12)^2}}{\left|2.24^2 + 7.94\left(0.81^2 - 0.4658^2\right)\right|} = 0.5301$$

$$C_p = \frac{7.94\left((0.81@104) - (0.4658@133.3)(0.57@-63)\right)}{2.24^2 + 7.94\left(.81^2 - 0.4658^2\right)} = 0.4471@102.3°$$

$$R_p = \frac{2.24\sqrt{2.24^2 - 2(7.94)(0.439)(2.24)(0.12) + 7.94^2(0.12)^2}}{\left|2.24^2 + 7.94\left(0.81^2 - .4658^2\right)\right|} = 0.7675$$

$$C_a = \frac{6.31\left((0.81@104) - \left(0.4658@133.3°\right)(0.57@-63)\right)}{2.24^2 + 6.31\left(0.81^2 - 0.4658^2\right)} = 0.4919@118.0°$$

$$R_a = \frac{2.24\sqrt{2.24^2 - 2(6.31)0.429(2.24)(0.12) + 6.31^2(0.12)^2}}{\left|2.24^2 + 6.31\left(0.81^2 - 0.4658^2\right)\right|} = 0.5825$$

$$C_p = \frac{6.31\left((0.81@104) - (0.4658@133.3)(0.57@-63)\right)}{2.24^2 + 6.31\left(.81^2 - 0.4658^2\right)} = 0.3662@102.3°$$

$$R_p = \frac{2.24\sqrt{2.24^2 - 2(6.31)(0.439)(2.24)(0.12) + 6.31^2(0.12)^2}}{\left|2.24^2 + 6.31\left(0.81^2 - .4658^2\right)\right|} = 0.7961$$

$$C_a = \frac{3.98\big((0.81@104) - (0.4658@133.3°)(0.57@-63)\big)}{2.24^2 + 3.98(0.81^2 - 0.4658^2)} = 0.3573@118.0°$$

$$R_a = \frac{2.24\sqrt{2.24^2 - 2(3.98)0.429(2.24)(0.12) + 3.98^2(0.12)^2}}{\left|2.24^2 + 3.98(0.81^2 - 0.4658^2)\right|} = 0.6870$$

$$C_p = \frac{3.98\big((0.81@104) - (0.4658@133.3)(0.57@-63)\big)}{2.24^2 + 3.98(.81^2 - 0.4658^2)} = 0.2417@102.3°$$

$$R_p = \frac{2.24\sqrt{2.24^2 - 2(3.98)(0.439)(2.24)(0.12) + 3.98^2(0.12)^2}}{\left|2.24^2 + 3.98(0.81^2 - .4658^2)\right|} = 0.8534$$

6.12

$$S_{11} = 0.56@140° \quad S_{21} = 1.74@32° \quad S_{12} = 0.15@48° \quad S_{22} = 0.41@-92°$$

$$\Delta = (0.56@140°)(0.41@-92°) - (1.74@32°)(0.15@48°) = 0.1386@-38.6°$$

$$K = \frac{1 - 0.56^2 - 0.41^2 + (0.1386)^2}{2(1.74)(0.15)} = 1.0297$$

$$G_{T\,max} = \frac{1.74}{0.15}\left(1.02970 - \sqrt{1.0297^2 - 1}\right) = 9.1093$$

Gains: 15.84, 12.59, 10, 7.94

$$C_a = \frac{7.94\big((0.81@104) - (0.4658@133.3°)(0.57@-63)\big)}{2.24^2 + 7.94(0.81^2 - 0.4658^2)} = 0.828@-145.8°$$

$$R_a = \frac{2.24\sqrt{2.24^2 - 2(7.94)0.429(2.24)(0.12) + 7.49^2(0.12)^2}}{\left|2.24^2 + 7.94(0.81^2 - 0.4658^2)\right|} = 0.137$$

Solutions

7.1

$$\omega_0 = \frac{2\pi(2.0\text{E}9 + 3.0\text{E}9)}{2} = 1.57\text{E}10$$

$$Q_L = \frac{1}{\omega_o RC} = 7.59$$

$$\omega' = \frac{2\pi(3.0\text{E}9 - 2.0\text{E}9)}{1.57\text{E}10} = 0.400$$

$$\left|\Gamma_{\min}\right| = \exp\left(-\frac{\pi(0.400)}{7.59}\right) = 0.8472$$

7.2

$$\omega_o = 2\pi(14.0\text{E}9) = 8.797\text{E}10$$

$$Q_L = 8.797\text{E}10(20)(0.48\text{E}-12) = 0.8445$$

$$\omega' = -\frac{Q_L \ln(0.1)}{\pi} = 0.6189$$

7.3

$\omega' = 0.667$

Input G_{max}:

$$Q_L = \frac{1}{2\pi(14E9)(2.5)(0.8E-12)} = 5.684$$

$$G_{max,in} = 1 - \exp\left(-\frac{\pi(0.667)}{5.684}\right) = 0.3082$$

Output G_{max}:

$$Q_L = 2\pi(14.0E9)(100)(1.6E-12) = 14.74$$

$$G_{max,out} = 1 - \exp\left(-\frac{\pi(0.667)}{14.74}\right) = 0.1383$$

Maximum gain:

$$G_{max} = 20(0.1383)*(0.3083) = 0.8523$$

7.4

$$\omega_0 = \frac{2\pi(2.0E9 + 12.0E9)}{2} = 4.40E10$$

$$\omega' = \frac{2\pi(12.0E9 - 2.0E9)}{4.40E10} = 1.429$$

Assume a 3 dB bandwidth.

$$\omega' = 2 - \frac{4}{\pi}\cos^{-1}\left|\frac{2(0.5)}{\ln\left(\frac{75}{40}\right)}\right|^{\frac{1}{N}} = 2 - \frac{4}{\pi}\cos^{-1}\left(1.591^{\frac{1}{N}}\right)$$

$$\cos\left(\frac{\pi(2-1.429)}{4}\right) = 1.591^{\frac{1}{N}} = 0.9010$$

$$N = \frac{\ln(1.591)}{\ln(0.9010)} = 4.45$$

$$N = 5$$

$$C_0^5 = \frac{5!}{5!0!} = 1$$

$$\ln\frac{Z_1}{Z_0} \approx 2^{-N} C_0^N \ln(R) = 2^{-5}(1)\ln\left(\frac{75}{40}\right) = 0.0196$$

$$\frac{Z_1}{Z_0} \approx \exp(0.0196) = 1.0198$$

$$C_1^5 = \frac{5!}{4!1!} = 5$$

$$\ln\frac{Z_2}{Z_1} \approx 2^{-N} C_1^N \ln(R) = 2^{-5}(5)\ln\left(\frac{75}{40}\right) = 0.0982$$

$$\frac{Z_2}{Z_1} \approx \exp(0.0982) = 1.1032$$

$$C_2^5 = \frac{5!}{3!2!} = 10$$

$$\ln\frac{Z_3}{Z_2} \approx 2^{-N} C_2^N \ln(R) = 2^{-5}(10)\ln\left(\frac{75}{40}\right) = 0.1964$$

$$\frac{Z_3}{Z_2} \approx \exp(0.1962) = 1.2171$$

$$C_3^5 = \frac{5!}{2!3!} = 10$$

$$\ln\frac{Z_4}{Z_3} \approx 2^{-N} C_3^N \ln(R) = 2^{-5}(10)\ln\left(\frac{75}{40}\right) = 0.1964$$

$$\frac{Z_4}{Z_3} \approx \exp(0.1962) = 1.2171$$

$$C_4^5 = \frac{5!}{1!4!} = 5$$

$$\ln \frac{Z_5}{Z_4} \approx 2^{-N} C_4^N \ln(R) = 2^{-5}(5) \ln\left(\frac{75}{40}\right) = 0.0982$$

$$\frac{Z_5}{Z_4} \approx \exp(0.0982) = 1.1032$$

$$C_5^5 = \frac{5!}{0!5!} = 1$$

$$Z_1 = 44.39\Omega$$

$$Z_2 = 48.97\Omega$$

$$Z_3 = 59.61\Omega$$

$$Z_4 = 65.76\Omega$$

$$Z_5 = 72.98\Omega$$

7.5

$$\omega_o = 2\pi(1.5\text{E}9) = 9.42\text{E}9$$

$$\omega' = \frac{(1.0\text{E}9 + 2.0\text{E}9)}{1.5\text{E}9} = 0.667$$

$$\omega_b = 9.42\text{E}9 - \frac{.667}{9.42\text{E}9} = 9.42\text{E}9$$

$$\sec\left(\frac{(9.42\text{E}9)\pi}{2(9.42\text{E}9)}\right) = \cos\left(\frac{1}{N} \cos^{-1}\left((0.8913)\frac{75-50}{75+50}\right)\right) = \cos\left(\frac{1.3916}{N}\right)$$

N = 2

Solutions

8.1

$$P_n = kTB = (1.374\text{E} - 23)(290) = 3.9846\text{E} - 21B$$
$$R = 10$$

$$\left\langle \left| e_n \right|^2 \right\rangle = 4kTRB = 4(1.374\text{E} - 23)(290)(10)B = 1.5938\text{E} - 19B \text{ volts}$$

$$\left\langle \left| i_n \right|^2 \right\rangle = 4kTGB = 4(1.374\text{E} - 23)(290)\left(\frac{1}{10}\right)B = 1.5938\text{E} - 21B \text{ volts}$$

$$R = 50$$

$$\left\langle \left| e_n \right|^2 \right\rangle = 4kTRB = 4(1.374\text{E} - 23)(290)(50)B = 7.9692\text{E} - 19B \text{ volts}$$

$$\left\langle \left| i_n \right|^2 \right\rangle = 4kTGB = 4(1.374\text{E} - 23)(290)\left(\frac{1}{50}\right)B = 3.1877\text{E} - 22B \text{ volts}$$

$$R = 350$$

$$\left\langle \left| e_n \right|^2 \right\rangle = 4kTRB = 4(1.374\text{E} - 23)(290)(350)B = 5.5784\text{E} - 18B \text{ volts}$$

$$\left\langle \left| i_n \right|^2 \right\rangle = 4kTGB = 4(1.374\text{E} - 23)(290)\left(\frac{1}{350}\right)B = 4.5538\text{E} - 23B \text{ volts}$$

$$R = 10,000$$

$$\left\langle \left| e_n \right|^2 \right\rangle = 4kTRB = 4(1.374\text{E}-23)(290)(10{,}000)B = 1.5938\text{E}-16B \text{ volts}$$

$$\left\langle \left| i_n \right|^2 \right\rangle = 4kTGB = 4(1.374\text{E}-23)(290)\left(\frac{1}{10000}\right)B = 1.5938\text{E}-24B \text{ volts}$$

8.2

$$I_1 = Y_{11}V_1 + Y_{21}V_2 + i_{n1}$$

$$I_2 = Y_{21}V_1 + Y_{22}V_2 + i_{n2}$$

Define an e_i such that

$$e_i = \frac{i_{n2}}{Y_{21}}$$

$$I_2 = Y_{21}V_1 + Y_{22}V_2 + e_i Y_{21}$$

$$I_2 = Y_{21}(V_1 - e_i) + Y_{22}V_2$$

Define an i_i such that

$$i_i = i_{n1} - Y_{11}e_i$$

$$I_1 = Y_{11}(V_1 - e_i) + Y_{21}V_2 + i_i$$

8.3

$$V_1 = Z_{11}(I_1 - i_i) + Z_{21}I_2 + e_i$$

$$V_2 = Z_{21}(I_1 - i_i) + Z_{22}I_2$$

Define an e_{n2} such that

$$e_{n2} = Z_{21}i_i$$

$$i_i = \frac{e_{n2}}{Z_{21}}$$

$$V_1 = Z_{11}\left(I_1 - \frac{e_{n2}}{Z_{21}}\right) + Z_{21}I_2 + e_i$$

$$V_2 = Z_{21}I_1 + Z_{22}I_2 + e_{n2}$$

Define an e_{n1} such that

$$e_{n1} = e_i - Z_{11}\frac{e_{n2}}{Z_{21}}$$

$$e_{n2} = \frac{Z_{21}}{Z_{11}}\left(e_i - e_{n1}\right)$$

$$V_1 = Z_{11}\left(I_1 - \frac{\frac{Z_{21}}{Z_{11}}\left(e_i - e_{n1}\right)}{Z_{21}}\right) + Z_{21}I_2 + e_i$$

$$V_1 = Z_{11}I_1 + Z_{21}I_2 + e_{n1}$$

8.4

$$P_n = kTB = (1.374\text{E} - 23)(70) = 9.618\text{E} - 22B$$

$$P_n = kTB = (1.374\text{E} - 23)(4) = 5.496\text{E} - 23B$$

8.5

Noise generated in the load is a_{nL}.

$$\begin{bmatrix} b_1 - b_n \\ b_2 \end{bmatrix} = \begin{bmatrix} S_{11} & S_{12} \\ S_{21} & S_{22} \end{bmatrix}\begin{bmatrix} a_1 + a_n \\ a_2 \end{bmatrix}$$

$$b_1 - b_n = S_{11}\left(a_1 + a_n\right) + S_{12}\left(a_2 + a_{nL}\right)$$

$$b_2 = S_{21}(a_1 + a_n) + S_{22}(a_2 + a_{nL})$$

$$a_2 = \Gamma_L b_2$$

$$b_1 - b_n = S_{11}(a_1 + a_n) + S_{12}\Gamma_L b_2 + S_{12}a_{nL}$$

$$b_2 = S_{21}(a_1 + a_n) + S_{22}\Gamma_L b_2 + S_{22}a_{nL}$$

$$b_2(1 - S_{22}\Gamma_L) = S_{21}(a_1 + a_n) + S_{22}a_{nL}$$

$$b_2 = \frac{S_{21}(a_1 + a_n) + S_{22}a_{nL}}{(1 - S_{22}\Gamma_L)}$$

$$b_1 - b_n = S_{11}(a_1 + a_n) + S_{12}\Gamma_L \frac{S_{21}(a_1 + a_n) + S_{22}a_{nL}}{(1 - S_{22}\Gamma_L)}$$

$$b_1 - b_n = \left(S_{11} + \frac{S_{21}S_{12}\Gamma_L}{(1 - S_{22}\Gamma_L)}\right)(a_1 + a_n) + \frac{S_{12}\Gamma_L S_{22}a_{nL}}{(1 - S_{22}\Gamma_L)}$$

When $a_1 = 0$, $b_1 = 0$ and

$$b_n = a_n\left(S_{11} + \frac{S_{21}S_{12}\Gamma_L}{(1 - S_{22}\Gamma_L)}\right) + a_{nL}\frac{S_{12}\Gamma_L S_{22}}{(1 - S_{22}\Gamma_L)}$$

8.6

$$F_{min} = 10^{0.23} = 1.698$$

$$F = 1.698 + \frac{4\ (2.3)}{50}\frac{(0.6708)^2}{1.432^2} = 1.7582 = 2.45\text{dB}$$

8.7

$$F_{min} = 10^{0.23} = 1.698$$

$$r_n = \frac{2.3}{50} = .046$$

$$\left|\Gamma_{opt}\right| = 0.6708$$

$$T_a = (1.698 - 1)290 + \frac{4\ (0.046)(290)(0.6708)^2}{1.4318^2} = 219.88$$

$$T_b = \frac{4\ (.046)(290)}{1.4318^2} - (1.698 - 1)290 = -176.39$$

$$T_c = \frac{4\ (0.046)\ (290)0.6708}{1.4318^2} = 174.60$$

$$\phi_c = \pi - 1.07149 = 2.0701r = 118.6°$$

$$T = \frac{T_a + 2\ \mathrm{Re}\{\Gamma_s\ T_c e^{j\phi_c}\} + T_b|\Gamma_s|^2}{\left(1 - |\Gamma_s|^2\right)}$$

8.8

$$\mathrm{ENR} = 10\ \log\left(\frac{2400 - 290}{290}\right) = 8.62\mathrm{dB}$$

8.9

Noise power at the input of the receiver:

$$P_a = kTB = (1.374\mathrm{E} - 23)(350)(4.8\mathrm{E}6) = 2.308\mathrm{E} - 14\mathrm{W}$$

Noise temperature of the receiver:

$$F = 10^{0.25} = 1.778$$

$$T = 290(1.778 - 1) = 225.7°$$

$$P_r = kTB = (1.374\mathrm{E} - 23)(225.7)(4.8\mathrm{E}6) = 1.489\mathrm{E} - 14\mathrm{W}$$

Power output from the receiver:

$$P_{out} = (2.308\mathrm{E} - 14 + 1.489\mathrm{E} - 14)\left(10^{4.6}\right) = 1.511\mathrm{E} - 9\mathrm{W}$$

8.10

Noise temperature of the receiver:

$$F = 10^{0.25} = 1.778$$

$$T = 290(1.778 - 1) = 225.7°$$

Noise source on:

$$P_{in} = kTB = (1.374\text{E} - 23)(225.7 + 290)(4.8\text{E}6) = 3.4011\text{E} - 14\text{W}$$

Power output from the receiver:

$$P_{out} = (3.4011\text{E} - 14)(10^{4.6}) = 1.354\text{E} - 9\text{W}$$

Noise temperature of the source when on:

$$290(10^{1.34} + 1) = 6634.5°$$

$$P_{in} = kTB = (1.374\text{E} - 23)(225.7 + 6634.5)(4.8\text{E}6) = 4.5244\text{E} - 13\text{W}$$

$$P_{out} = (4.5244\text{E} - 13)(10^{4.6}) = 1.801\text{E} - 8\text{W}$$

9

Solutions

9.1

12 GHz	0.70 dB	0.43	165.2°	2 Ω	10 dB

$$F_{\min} = 10^{0.07} = 1.1749$$

$$F = 10^{0.085} = 1.2162$$

$$N = \frac{50(1.2162 - 1.1749)(0.5945)^2}{4(2)} = 0.0912$$

$$C_f = \frac{0.43@165.2}{0.0912 + 1} = 0.394@165.2°$$

$$R_f = \frac{\sqrt{0.0912\left(0.0912 + \left(1 - (0.43)^2\right)\right)}}{0.0912 + 1} = 0.3126$$

$$F = 10^{0.10} = 1.2589$$

$$N = \frac{50(1.2162 - 1.1749)(0.5945)^2}{4(2)} = 0.1856$$

$$C_f = \frac{0.43@165.2}{0.1856+1} = 0.3627@165.2°$$

$$R_f = \frac{\sqrt{0.1856\left(0.1856+\left(1-(0.43)^2\right)\right)}}{0.1856+1} = 0.4254$$

$$F = 10^{0.14} = 1.3804$$

$$N = \frac{50(1.2162-1.1749)(0.5945)^2}{4(2)} = 0.1856$$

$$C_f = \frac{0.43@165.2}{0.4539+1} = 0.2958@165.2°$$

$$R_f = \frac{\sqrt{0.4539\left(0.4539+\left(1-(0.43)^2\right)\right)}}{0.4539+1} = 0.5932$$

9.2

$S_{11} = 0.863@{-}79.1°$ $S_{12} = .072@36.5°$ $S_{21} = 3.434@106.2°$ $S_{22} = 0.627@{-}58.3°$

$F_{min} = 0.46$ dB $R_n = 13.5$ $\Gamma_{opt} = 0.740@62.0°$

$$F_{min} = 10^{0.048} = 1.1169$$

$$F = 10^{0.125} = 1.3335$$

$$N = \frac{50(1.333-1.1169)(1.4975)^2}{4(13.5)} = 0.0090$$

$$C_f = \frac{0.740@62°}{0.0090+1} = 0.7334@62°$$

$$R_f = \frac{\sqrt{0.0090\left(0.0090+\left(1-(0.740)^2\right)\right)}}{0.0090+1} = 0.1173$$

$$G = 10^{1.55} = 35.41$$

$K = 0.3420$

Source stability circle:

$C_s = 1.3059 @ 94.6°$

$R_s = .5648$

Constant Gain circle:

$C_a = 0.7423 @ 94.54°$

$R_a = 0.4411$

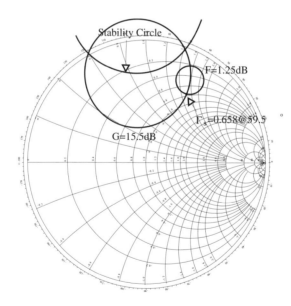

$\Gamma_s = 0.658 @ 59.5°$

$\Gamma_0 = S_{22} + \dfrac{S_{21}S_{12}\Gamma_s}{1 - S_{11}\Gamma_s} = 0.5331 @ -89.4°$

9.3

$S_{11} = 0.56@96°$ $S_{12} = 0.131@-29°$ $S_{21} = 1.58@-33°$ $S_{22} = 0.41@-166°$

$F_{min} = 2.15$ dB $R_n = 12.5$ $\Gamma_{opt} = 0.41@-152.0°$

$F_{min} = 10^{0.048} = 1.6406$

$F = 10^{0.25} = 1.783$

$N = 0.0611$

$C_f = \dfrac{0.740@62°}{0.0090+1} = 0.3864@-152°$

$R_f = 0.2584$

$K = 1.258$

$G_{T\,max} = 5.9671$

$\Gamma_{sM} = .7532@-97.7°$

Constant Gain circle:

$G = 5.8$
$C_a = 0.7407@-97.7°$
$R_a = 0.0857$
$G = 5.5$
$C_a = 0.7180@-97.7°$
$R_a = 0.1466$
$G = 5.0$
$C_a = 0.6778@-97.7°$
$R_a = 0.2214$
$G = 5.0$
$C_a = 0.6343@-97.7°$
$R_a = 0.2871$

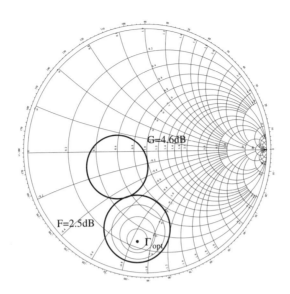

$$\Gamma_s = 0.4158 @-115$$

$$\Gamma_0 = S_{22} + \frac{S_{21}S_{12}\Gamma_s}{1-S_{11}\Gamma_s} = 0.5174 @-168.9°$$

9.4

$$S_{11} = 0.86@-59° \quad S_{12} = 0.101@64° \quad S_{21} = 2.27@120° \quad S_{22} = 0.57@-17°$$

$$F_{min} = 1.1 \text{ dB} \quad R_n = 54 \quad \Gamma_{opt} = 0.67@52°$$

$$F_{min} = 10^{0.11} = 1.2882$$

$$K = 0.5831$$

Stability circles:

$$C_s = 1.404 @72.0°$$
$$R_s = 0.5623$$
$$C_L = 32.2138@-103.5°$$
$$R_L = 32.787$$

Lowest noise figure is at Γ_{opt}.

$$\Gamma_0 = S_{22} + \frac{S_{21}S_{12}\Gamma_s}{1 - S_{11}\Gamma_s} = 0.5209@{-}54°$$

$$\Gamma_L = 0.5209@54.5°$$

This is in the stable region of the load.

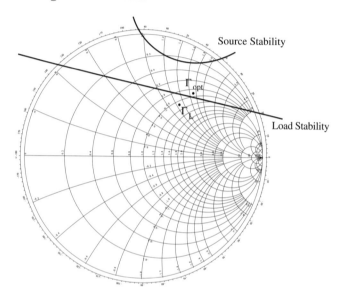

9.5

	Amplifier 1	Mixer	Filter	Amplifier 2
Gain	10	0.25	0.40	1585
Noise Factor	1.41	6.31	2.51	2.51

$$F_t = 1.41 + \frac{6.31-1}{10} + \frac{2.51-1}{(10)(0.25)} + \frac{2.51-1}{(10)(0.25)(0.4)} = 4.0550 = 6.08\text{dB}$$

	Amplifier 1	Mixer	Filter	Amplifier 2
Gain	100	0.25	0.40	1585
Noise Factor	1.41	6.31	2.51	2.51

$$F_t = 1.41 + \frac{6.31 - 1}{100} + \frac{2.51 - 1}{(100)(0.25)} + \frac{2.51 - 1}{(100)(0.25)(0.4)} = 1.6745 = 2.24\text{dB}$$

9.6

$$\text{MSG} = 19\text{dB} = 79.43$$

$$NM = \frac{(79.43)(1.11) - 1}{79.43 - 1} = 1.111$$

9.7

	Amplifier 1	Mixer	Filter	Amplifier 2
Gain	31.62	0.25	0.40	1585
IP3	100	10	1e100	1000

$$\frac{1}{IP3_{total}} = \frac{1}{1000} + \frac{1}{1E1000(1585)} + \frac{1}{10(1585)(0.25)} + \frac{1}{100(1585)(0.25)(0.4)} = .0013$$

$$IP3_{total} = 794.5 = 29.0\text{dB}$$